I0797535

US MILITARY CAREERS

US AIR FORCE

BY CARLA MOONEY

CONTENT CONSULTANT
Robert Wettemann Jr., PhD
Associate Professor
Department of History
United States Air Force Academy

An Imprint of Abdo Publishing | abdobooks.com

ABDOBOOKS.COM

Published by Abdo Publishing, a division of ABDO, PO Box 398166, Minneapolis, Minnesota 55439.

Printed in the United States of America, North Mankato, Minnesota.
042020
092020

Cover Photo: Kemberly Groue/US Air Force
Interior Photos: Senior Airman Kayla Newman/US Air Force/Defense Visual Information Distribution Service, 4–5; iStockphoto, 9; John T. Daniels/Virginian–Pilot/AP Images, 12–13; AP Images, 14; Library of Congress, 16; FS/AP Images, 20; US Air Force, 23; Airman 1st Class Bryan Guthrie/US Air Force/Defense Visual Information Distribution Service, 25; Master Sgt. Joshua L. DeMotts/US Air Force/Defense Visual Information Distribution Service, 26–27; Tech. Sgt. Liliana Moreno/US Air Force/Defense Visual Information Distribution Service, 31; T. Whitney/Shutterstock Images, 33; Benjamin Faske/US Air Force, 36; Tech. Sgt. Michael Charles/US Air Force/Defense Visual Information Distribution Service, 38–39; Joshua J. Seybert/US Air Force/Defense Visual Information Distribution Service, 40; James M. Bowman/US Air Force, 43; Airman 1st Class Hannah Strobel/US Air Force/Defense Visual Information Distribution Service, 48; Airman 1st Class Dwane R. Young/US Air Force/Defense Visual Information Distribution Service, 50–51; Airman 1st Class Bailee A. Darbasie/US Air Force/Defense Visual Information Distribution Service, 54; Senior Airman Christopher Maldonado/US Air Force/Defense Visual Information Distribution Service, 56; Senior Airman Malcolm Mayfield/US Air Force/Defense Visual Information Distribution Service, 60–61; Ismael Ortega/US Air Force/Defense Visual Information Distribution Service, 63; Alejandro Peña/US Air Force/Defense Visual Information Distribution Service, 69; Airman 1st Class Eugene Oliver/US Air Force/Defense Visual Information Distribution Service, 72–73; Senior Airman Dillon J. Audit/US Air Force/Defense Visual Information Distribution Service, 75; Senior Airman Abby L. Finkel/US Air Force/Defense Visual Information Distribution Service, 77; Airman 1st Class Seth Haddix/US Air Force/Defense Visual Information Distribution Service, 80–81; Tech. Sgt. Terrica Y. Jones/US Air Force/Defense Visual Information Distribution Service, 84; Staff Sgt. Matthew J. Wisher/US Air Force/Defense Visual Information Distribution Service, 88–89; Staff Sgt. Anthony Agosti/US Air National Guard/Defense Visual Information Distribution Service, 93; Kemberly Groue/US Air Force/Defense Visual Information Distribution Service, 96

Editor: Charly Haley
Series Designer: Nikki Nordby

LIBRARY OF CONGRESS CONTROL NUMBER: 2019954355

PUBLISHER'S CATALOGING-IN-PUBLICATION DATA

Names: Mooney, Carla, author.
Title: US Air Force / by Carla Mooney
Description: Minneapolis, Minnesota : Abdo Publishing, 2021 | Series: US military careers | Includes online resources and index.
Identifiers: ISBN 9781532192265 (lib. bdg.) | ISBN 9781098210168 (ebook)
Subjects: LCSH: United States. Army Air Forces--Juvenile literature. | Air force personnel--Juvenile literature. | Military power--Juvenile literature. | United States. Air Force--History--Juvenile literature. | Armed Forces--Juvenile literature.
Classification: DDC 355.12--dc23

CONTENTS

CHAPTER 1

FLYING BEHIND ENEMY LINES

On a dark night, a helicopter took off from an American air base in Uzbekistan. Once in the air, the copter turned south and flew into Afghanistan on a secret mission. It carried one of the US Air Force's most elite and little-known special operators. The airman had been carefully selected for this

A US military helicopter flies into the night.

important mission in the War on Terror. If he was successful, he would not have to shoot his gun. No one outside of his military team would know what he had done. He was a special operations weather technician (SOWT).

MISSION-CRITICAL WEATHER FORECAST

As commando weather forecasters, US Air Force SOWTs gather mission-critical environmental data from some of the most difficult and hostile places on the planet. They work in special operations units with Navy SEALs, Delta Force, and Army Rangers. When a major military operation is planned, SOWTs go on the ground to give a weather forecast that will determine whether the operation begins as planned or is put on hold.

On this night, the helicopter carried the air force SOWT to a strip of desert about 80 miles (129 km) south of Kandahar, Afghanistan. The weather officer had a satellite forecast that called for clear skies that night. Yet the air force also needed data from on the ground, and there was no weather information coming from Afghanistan. To gather that critical data, the SOWT prepared to jump behind enemy lines.

SPECIAL OPERATIONS FORCES

The US military has many tough soldiers, airmen, and sailors. Yet within the service branches, there are elite warriors who take on the toughest assignments. For the US Air Force, that is the Air Force Special Operations Command, established in 1990. The air force had other forms of special operations forces before that. Across the military, special operations forces embark on counterterrorism missions, direct raids and ambushes, hostage rescue and recovery missions, special reconnaissance behind enemy lines, and more. To successfully carry out these dangerous missions, the members of the Army Rangers, Navy SEALs, Air Force Special Ops, Army Green Berets, and other special operations units must all be in top physical and mental condition. To identify those who can meet the challenge of special operations, recruits must endure intense and grueling training that is designed to identify only the best people for the job.

As the helicopter neared the drop zone, the SOWT watched through night-vision goggles as a sandstorm swirled below. The pilot pulled the helicopter into a hover, and the SOWT descended 60 feet (18 m) by rope to the ground. A small team of air force combat commandos joined him. By dawn, the small team had traveled across several miles of desert, climbed a mountain, and set up on a ledge. There, the SOWT began his work.

For the next few days, he gathered information. He used laser range finders to determine cloud height. At night, he launched weather balloons to gather data from the upper atmosphere. He used a handheld device to gather additional weather data. With this data, the SOWT prepared a daily forecast, which he compared against computer weather

ENLISTED AIRMEN

Most people accepted into the US Air Force are enlisted airmen. Enlisted airmen must have a high school diploma or a GED. They are the general workforce of the air force, carrying out the air force's daily operations. They are often highly specialized and perform many hands-on tasks. Enlisted airmen can work in hundreds of jobs in dozens of career fields, from special operations to logistics and transportation. Joining the air force as an enlisted airman generally requires an overall service commitment of eight years, with either four or six years of that being served in active duty. The remaining commitment is served as a member of the inactive ready reserves, which are service members who are ready to be called upon to replace active duty airmen if needed. Enlisted airmen cannot quit until they have fulfilled their service commitment.

predictions. He adjusted the forecasts, added data, and recalculated the forecasts. He searched for forecasts with clear skies, moderate temperatures, calm winds, and good visibility.

By his third day in enemy territory, the SOWT found the right conditions. He wrote "Conditions Favorable" in a secure text message to the general who headed the US military's Joint Special Operations Command in Uzbekistan. "Roger," the general replied. "Force will launch." The Army Rangers waiting for the go-ahead signal launched their operation.

DEFENDING THE AIR

For more than a century, the US Air Force has defended the United States. It defends the country in the air. Air force personnel are trained and equipped to fly and fight to win wars and deter threats. They train for nearly every scenario, from terrorist attacks to catastrophe relief. In addition to defending

COMMISSIONED OFFICERS

In the air force, some personnel are commissioned officers who have a minimum of a four-year degree from a college or university and have completed officer training. Some are graduates of the US Air Force Academy, which prepares recruits to become officers. Commissioned officers lead and manage enlisted airmen. There are jobs available to officers in many career fields, from health and medicine to engineering and applied science. Air force officers usually have a service commitment of four years, with some positions, such as navigators and pilots, requiring longer service commitments.

★ The air force uses many different aircraft, along with other technology and equipment.

the United States and its citizens through the air, the air force also provides humanitarian aid when disasters strike around the world.

Many people are needed to keep the US Air Force operating smoothly. The air force employs approximately 328,000 active duty troops, 68,000 reserve personnel, and 106,000 personnel

AIR NATIONAL GUARD

The Air National Guard is a reserve component of the US Air Force. Members train part time, close to home. They serve both state and federal governments. At the state level, Air National Guard units are the militia air force of each state. Directed by the state's governor, the Air National Guard assists communities during times of need. When needed, Air National Guard units can be federalized by order of the US president. When this occurs, Air National Guard units become an active piece of the US Air Force. There are more than 140 Air National Guard units in the United States and its territories.[2]

in the Air National Guard.[1] When joining the US Air Force, recruits can choose to enter active duty, where they serve as full-time air force service members. Alternately, they may choose to join the air force reserves and serve on a part-time basis. Air force reservists generally train one weekend per month and two full weeks per year while working in a civilian career. When needed, reservists may be called up to active duty. Air National Guard members follow a part-time training and work schedule similar to that of reservists.

Air force personnel operate, maintain, and repair the air force's aircraft. They live and work on air force bases around the world. Each year, thousands of recruits join the air force, filling new positions and jobs left open by service members who have returned to civilian life.

The US Air Force has numerous career fields. Service members can choose careers in logistics or as part of the aircrew operations team. They can work as pilots, in intelligence, or in cyberspace support. With so many opportunities, a career in the air force can fit many backgrounds and interests.

CHAPTER 2

THE HISTORY OF THE US AIR FORCE

In December 1903, Wilbur and Orville Wright successfully flew the first self-propelled airplane near Kitty Hawk, North Carolina. The US Army took notice of this new and exciting technology. With that first flight, the age of aviation had officially

Orville Wright pilots an airplane in 1903 as his brother, Wilbur Wright, watches. The Wright brothers paved the way for aviation.

begun. As flight technology continued to develop, it would eventually push the US military to create the US Air Force.

THE ARMY'S 1ST AERO SQUADRON

To explore the potential for aircraft and flight in the military, the US Army Signal Corps created a small aeronautical division on August 1, 1907. The division was tasked with studying and implementing the use of aircraft. In 1908, the Signal Corps began testing the army's first airplane in Fort Myer, Virginia.

★ The Wright brothers demonstrate one of the US military's first airplanes in Virginia in 1909.

However, the plane crashed during a test flight. In 1909, the Signal Corps replaced the plane with an improved Wright Flyer, developed by the Wright brothers. It became known as Airplane No. 1.

In 1913, army aviators formed the 1st Aero Squadron. It was the first military unit of the US Army that was specifically focused on aviation. In 1916, the unit participated in a military operation on the Mexican border, becoming the US Army's first air combat unit. Since its creation, this unit has been continuously active, operating today as the 1st Reconnaissance Squadron.

In 1914, the US Congress created an Aviation Section within the US Army Signal Corps, replacing the original aeronautical division. A new law authorized the Aviation Section to operate and supervise all army aircraft, including hot air balloons and airplanes, along with the related equipment. The Aviation Section would also train officers and enlisted soldiers in military aviation.

WORLD WAR I

In August 1914, World War I (1914–1918) began in Europe. In previous wars, fighting took place on battlefields on the ground and with warships at sea. In World War I, a new battle front emerged—the skies. At first, aircraft were used for reconnaissance and to see what was happening on the ground. As pilots from each side encountered each other, they began to engage by throwing objects and firing small arms. Over time, the aircraft began to carry machine guns and engage in aerial combat. Airplanes attacked troops on the ground and engaged in midair fighting.

At the time, the 1st Aero Squadron was the US Army's only aviation unit. It was a tiny air force compared with the airpower of the European countries. The American unit had only 12 officers, 54 enlisted men, and six aircraft.[1] The war in Europe caused the American military to plan an expansion of its air force.

★ A pilot stands in front of a US Army airplane during World War I.

Before the United States officially joined the war, some of the first Americans to fight in Europe did so as part of the Lafayette Escadrille. On March 21, 1916, they were authorized to fly under the protection of French air forces over the war's Western front. By the time the United States entered World War I in 1917, its air force had grown. The Army Signal Corps' Aviation Section became the US Army Air Service. At its peak during the war, the Air Service commanded more than 7,800 planes.[2] At first, control of the aero squadrons was spread among various army organizations, which made it difficult to coordinate air missions. To resolve this, aero squadrons with similar functions were grouped together.

By the end of the war in 1918, the Air Service had expanded to 185 aero squadrons, along with 86 balloon companies and numerous other units. More than 178,000 enlisted soldiers and 19,000 officers served in the army's Air Service.[3]

PEACETIME ACTIVITIES

After World War I, the Air Service quickly decreased in size. In 1920, the Army Reorganization Act made the Air Service a combatant arm of the army, which meant that it could be part of direct combat missions. The Air Service also developed formal training for its service members. It brought pilots to Texas for flight training. The Air Service also established technical schools for both officers and enlisted soldiers. Officers also attended tactical school, which trained them to lead military aviation units. In 1926, the name of the Air Service was changed to the Air Corps, but it remained part of the US Army.

Recognizing the growing importance of military airpower, President Franklin D. Roosevelt formally requested that the US Congress authorize an expansion of the Air Corps in 1939.

As war threatened again in Europe, Congress authorized $300 million for the Air Corps expansion and production of airplanes.

WORLD WAR II

In September 1939, Germany invaded Poland, and World War II (1939–1945) began in Europe. Within a year, the Germans had occupied Poland, Norway, the Netherlands, Belgium, and France. World War II engaged more than 30 countries in the deadliest conflict in the world's history. The battlefield in the sky became crucial to victory. Great Britain's Royal Air Force (RAF) battled Germany's Luftwaffe for control of the skies over Britain and Europe. During the Battle of Britain, the Royal Air Force organized three fighter squadrons

TUSKEGEE AIRMEN

Before and during World War II, the US military was mostly segregated by race. Black airmen primarily served as enlisted soldiers in support roles. When World War II broke out in Europe, President Franklin D. Roosevelt ordered the US Army to begin training black pilots at Tuskegee Army Air Field in Alabama. At the time, many white Americans believed that black people were not skilled enough to fly a plane or use highly technical equipment. However, the Tuskegee Airmen proved them wrong. They risked their lives and flew approximately 15,000 missions in Europe and North Africa during World War II.[4] They defeated hundreds of enemy planes and sank an enemy ship. The success of these brave pilots was one of the factors that pushed the US military to stop segregating its personnel after the war.

of volunteer American pilots to fly with the RAF. Known as the heralded "Eagle Squadrons," many of the American pilots continued to serve in the US military after the United States entered the war. Although the United States was not officially in the war right away, the US Army began to prepare and increased its number of personnel and aircraft.

On December 7, 1941, the Japanese military bombed the US Navy base in Pearl Harbor, Hawaii, and the United States entered World War II. The United States joined the Allied forces with Britain, France, and Russia to battle against the Axis Powers of Germany, Italy, and Japan. The Air Corps had a critical role in World War II. Its efforts were important to Allied victories in both the Atlantic and Pacific regions. In Europe, the Air Corps' bombers embarked on an intense campaign to crush German defenses and prepare Europe for an Allied invasion. By the war's end, the German Luftwaffe struggled to match the overwhelming airpower of the US Army Air Corps, now called the US Army Air Forces.

In the Pacific, the air force fought alongside the US Navy in many battles. As the US military moved closer to mainland Japan, Army Air Forces bombers pounded Japan with firebombs. The bombing campaign crippled Japanese factories. Then, in August 1945, B-29 bombers of the US Army Air Forces

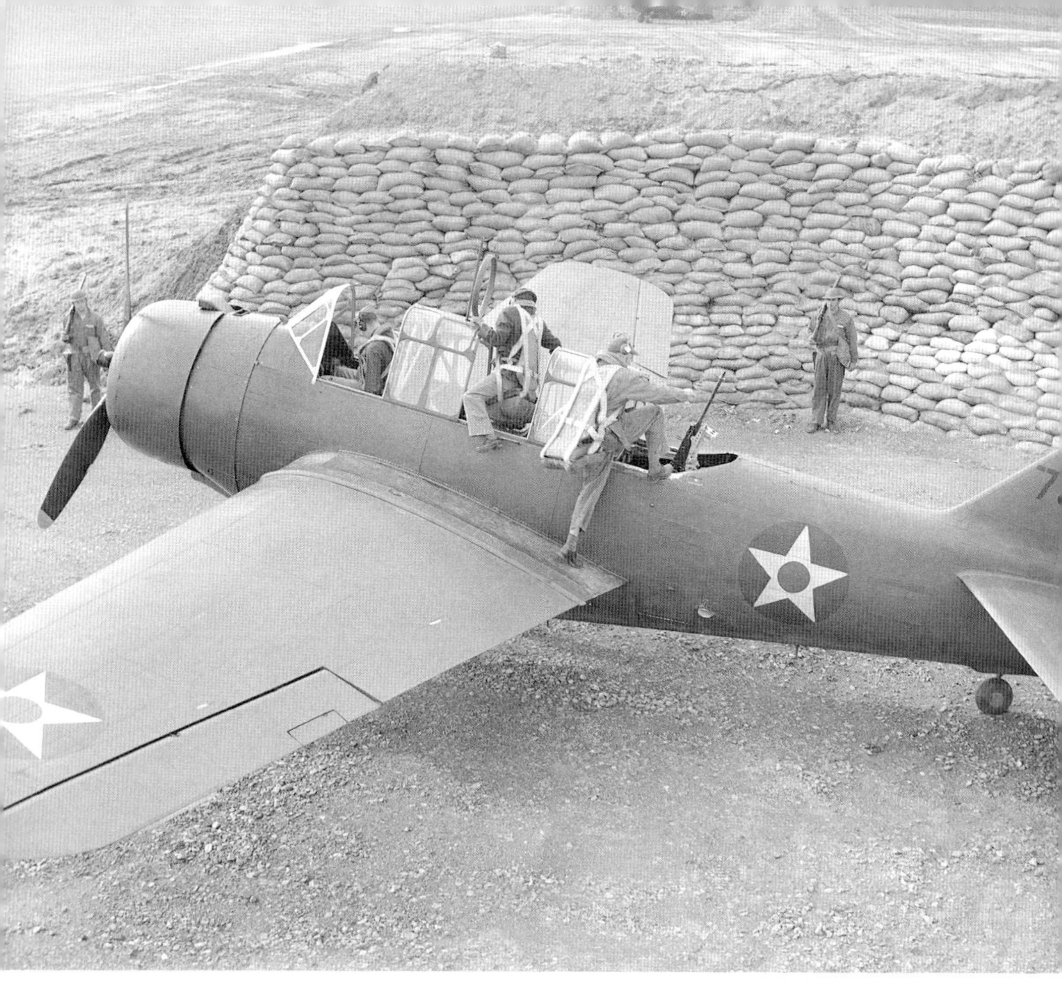

★ A crew boards a US Army observation airplane in 1942, during World War II.

dropped atomic bombs on the Japanese cities of Hiroshima and Nagasaki, which ended the war in the Pacific.

By the end of World War II, the US Army's air force had become a major military organization. It was made of up of numerous commands, divisions, wings, groups, squadrons, and other units. When the war ended in 1945, the US Army Air Forces included 2.2 million men and women and more than 63,000 aircraft.[5]

A SEPARATE AIR FORCE

The nation's air force had grown so large, it was ready to become an independent branch of the US military. The National Security Act of 1947 established the US Air Force as its own branch of the US military. It also established the Department of the Air Force, led by the Secretary of the Air Force. The Department of the Air Force is under the US Department of Defense, along with the Department of the Army and the Department of the Navy.

After World War II, the United States was considered a global superpower. In the following years, the US Air Force developed increasingly sophisticated weapons and technology. It displayed some of these achievements during the Korean War (1950–1953). The US Air Force flew jet fighter combat missions, aerial medical evacuations, and aerial resupply missions.

COLD WAR

After World War II, the relationship between the United States and the Soviet Union deteriorated. The two countries entered an arms race. Each nation attempted to increase its influence throughout the world. They also raced to develop new technologies and weapons. This period was known as the Cold War (1947–1991). During these years, the United States expanded its military forces around the world. The US Air Force opened air bases in Europe, Japan, and South Korea. During the Cold War, the air force was also responsible for the country's nuclear weapons carried by ground-launched intercontinental missiles and bomber aircraft, though none of these weapons were used in combat.

In the decade after the Korean War, the air force continued to develop its technology and capabilities. Airmen flew faster, higher, and longer than ever before. In the late 1960s, the United States was again drawn into conflict in Asia to halt the spread of communism. In the Vietnam War (1954–1975), the US Air Force fought with air-to-air missiles, performed search-and-rescue missions, and engaged in massive bombing operations. This included Operation Rolling Thunder, which lasted from 1965 to 1968. In that time, US Air Force planes conducted more than 153,000 attacks against the communist North Vietnam. A total of 864,000 tons (783,800 metric tons) of American bombs were dropped on North Vietnam during Rolling Thunder.[6] With its use of new tactics and sophisticated weapons systems, the air force changed the way wars would be fought in the future.

Outside of combat, the air force also developed new technologies for the country's defense. This included early warning networks for aircraft (which allowed the military to detect enemy aircraft earlier) and intercontinental ballistic missiles. The air force also pioneered many of the global positioning satellites used today. This technology allowed the military to become more precise by using global positioning system (GPS) coordinates to guide weapons. The air force also developed rocketry and spacecraft technology that laid the foundation for

★ US Air Force planes fly over burning oil fields in Kuwait during Operation Desert Storm.

the United States putting satellites into orbit around Earth and landing astronauts on the moon.

OPERATION DESERT STORM

In 1990, Iraq invaded the neighboring country of Kuwait, beginning the Gulf War (1990–1991). The invasion was condemned by the United States, Britain, and the Soviet Union. The United Nations Security Council authorized the use of all necessary force if Iraq did not withdraw from Kuwait. When Iraq refused, the US Air Force led a massive air operation known as Operation Desert Storm against the Iraqi military in January 1991. Along with coalition forces from other nations, the US Air Force used the latest military technology, including stealth bombers, cruise missiles, and smart bombs with laser-guidance systems, to attack and destroy the Iraqi air force.

For 42 days, the American-led forces relentlessly launched more than 100,000 air raids against the Iraqis and dropped more than 80,000 tons (72,600 metric tons) of military weapons and explosives.[7] The overwhelming power shown by the air forces led to the Gulf War's end in February 1991.

THE MODERN AIR FORCE

Today, the US Air Force continues to use innovative and sophisticated air and space technology to protect the country.

AIR FORCE SPACE COMMAND

In 1982, the Air Force Space Command (AFSPC) was formed. Its primary mission was space operations. The AFSPC grew to be responsible for launching, operating, and protecting satellites for military use. The military uses satellites to gather weather data, to enable communications, and for the GPS. During the ongoing War on Terror, the AFSPC provided space-based support to the US military in areas such as communications, global positioning, navigation, and meteorology. In December 2019, the AFSPC became the US Space Force, a separate branch of the military. In the future, the space force may develop combat aircraft that are able to enter space.

★ Some air force jobs require a lot of time at computers.

One of these technologies, remotely piloted aircraft (also called unmanned aerial vehicles or drones), has once again changed the way the air force fights battles. Using remotely piloted aircraft, the air force flies over dangerous areas to gather information or drop bombs while pilots remain in the safety of a ground control facility far from the battlefield.

Additionally, air force pilots fly fighters and bombers protected by stealth technology. Space operators launch satellites and sensors to watch enemies worldwide. As the world continues to evolve, the US Air Force aims to anticipate and meet changing global threats to keep the United States and its allies safe.

CHAPTER 3

THE US AIR FORCE TODAY

Today, the US Air Force is the largest and most technologically advanced air force in the world. It is one of the six branches of the US military. The others are the US Army, US Navy, US Marine Corps, US Coast Guard, and US Space Force. As of January 2020, the air force had 328,255 active duty

Two US airmen pose for a photo at an air base in Qatar. US Air Force personnel work all over the world.

personnel.[1] The air force is the branch of the military responsible for providing the United States with the ability to fly and fight battles in the air.

MAJOR COMMANDS

The US Air Force is organized in a tiered structure. At the top is the US president in his role as commander in chief of the entire US military, followed by the US Secretary of Defense. Then comes the Secretary of the Air Force (SECAF), a civilian who leads the Department of the Air Force. The SECAF works with the Chief of Staff of the Air Force, who is the highest-ranking military officer in the Department of the Air Force. They work on issues that affect the entire air force.

The air force's personnel are divided into 11 Major Commands (MAJCOMs). The MAJCOMs are typically led by three- or four-star generals. MAJCOMs located in the continental United States are organized by function. Overseas, they are organized by geography.

There are eight MAJCOMs in the continental United States. The Air Combat Command provides combat airpower, including fighters, bombers, and other combat aircraft.

MILITARY ENTRANCE PROCESSING STATIONS

When a person decides to join the US Air Force, the first step is a Military Entrance Processing Station (MEPS). All new members of the air force and other US military branches go through MEPS to ensure they meet the mental, moral, and medical standards required by the Department of Defense. Across the United States, there are 65 MEPS locations. At a MEPS location, applicants for each service take a career aptitude test. They also complete an extensive medical exam, which includes a weight check, hearing test, and vision test. Together with a service counselor, they select a military job and also participate in an interview that is designed to highlight any potential legal issues that would prevent them from enlisting in the US military. Once it is determined that the applicant is qualified to enlist, he or she takes the oath of enlistment. The new recruit then gets ready for basic training.

US AIR FORCE ACADEMY

As the air force grew during World War II, airpower leaders argued that they needed a military school dedicated to training air force officers, similar to the army's West Point or the navy's US Naval Academy. In 1954, the military established the US Air Force Academy. In 1955, construction of the new academy began in Colorado Springs, Colorado, and its first class of cadets graduated in 1959. In 1976, the academy began allowing women to enter as cadets. Before that, only men were allowed to enlist.

Today, the academy is home to approximately 4,000 cadets.[2] The cadets take a core set of classes and can choose from numerous major fields of study, along with a variety of elective classes and research opportunities. Upon graduation, students earn a bachelor of science degree. With state-of-the-art academic and professional training, the academy prepares its cadets to become officers to help lead the US Air Force.

The Air Education and Training Command handles all recruiting, training, and education of airmen. The Air Force Global Strike Command provides intercontinental ballistic missile forces and nuclear-capable bomber forces. The Air Force Materiel Command researches, develops, and tests weapons systems used by the air force. The Air Force Reserve Command manages the service's reserve units. The Air Force Space Command is responsible for providing space and cyberspace forces, as well as developing and testing space systems. The Air Force Special Operations Command provides special operations airpower to missions worldwide. The Air Mobility Command rapidly moves people, materials, and equipment to and from an area by air.

The air force has three MAJCOMs overseas. The Pacific Air Forces is based in Hawaii and provides airpower in the Pacific region. The US Air Forces in Europe–Air Forces Africa is based in Germany and serves as the air component for the US military's European and African operations, including combat, humanitarian, and peacekeeping missions. The US Air Forces Central Command provides air operations and power in Southwest Asia.

OTHER DIVISIONS

The MAJCOMs are divided into Numbered Air Forces (NAFs). NAFs are primarily used in wartime and are often assigned to provide air operations to a specific geographic area. Within each NAF, air force personnel are organized into wings, groups, and squadrons. Each wing contains about 1,000 to 5,000 personnel.[3] Some wings are operational and include all of the necessary support functions to operate independently. Other wings are assigned to an air base and conduct the base's operations, while some wings are assigned to specialized missions.

Most wings include approximately four groups. Typically, they consist of an operational group that includes pilots and crews, a maintenance group, a mission support group (which includes a variety of personnel in engineering, logistics, communications, and more), and a medical group. Within each group, there

★ Members of the air force's 821st Contingency Response Group line up at Travis Air Force Base in California.

are squadrons. Squadrons can include up to 24 aircraft and their crews. Each squadron has a specialized function. Some squadrons are fighter squadrons or bomber squadrons, while others are training squadrons or medical squadrons.[4] Each squadron can also be divided into flights for certain missions, with up to 100 personnel in each flight.[5]

AIR FORCE RANKS

Air force personnel generally fall into three main categories: enlisted airmen, noncommissioned officers, and commissioned officers. Enlisted airmen are the main workforce of the air force.

When a recruit joins the US Air Force as an enlisted airman, he or she starts at the rank of airman basic. After completing recruit training, the enlisted airman advances to the rank of airman. As enlisted airmen gain experience, they can earn promotions to higher ranks, which come with more pay. As enlisted airmen move up the ranks, they can become noncommissioned officers. The highest enlisted rank is chief master sergeant of the air force, a senior noncommissioned officer.

While enlisted military personnel do not need college degrees, commissioned officers must have at least a four-year degree and must complete officer training. The most junior commissioned officer in the air force is a second lieutenant. Experienced officers can earn promotions to higher ranks such as lieutenant or captain. There are ten officer ranks. The highest air force rank is general of the air force, used only during times of war.

AIRMAN SKILL LEVELS

When airmen first enter the air force, after they have finished basic military training, they are designated at skill level 1 (helper). When they graduate technical school, which is training for their specific career, they become skill level 3 (apprentice). After a period of 12 to 18 months of on-the-job training and online courses, airmen progress to skill level 5 (journeyman). Airmen who are promoted to staff sergeant enter training for skill level 7 (craftsman). At this level, airmen can participate in more online and on-the-job training courses. For some jobs, level-7 airmen attend specialized technical schools. When promoted further, airmen can receive skill level 9 (superintendent).

US AIR FORCE RANKS

ENLISTED AIRMEN

Airman Basic

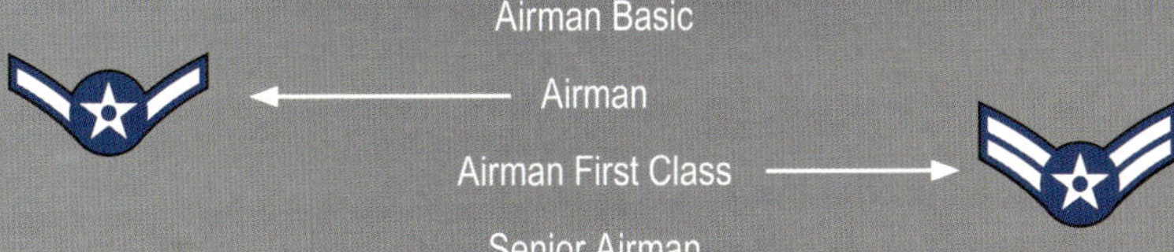

Airman

Airman First Class

Senior Airman

NONCOMMISSIONED OFFICERS (ENLISTED)

Staff Sergeant

Technical Sergeant

Master Sergeant/First Sergeant

Senior Master Sergeant/First Sergeant

Chief Master Sergeant/First Sergeant

Command Chief Master Sergeant

Chief Master Sergeant of the Air Force

COMMISSIONED OFFICERS

Second Lieutenant

First Lieutenant

Captain

Major

Lieutenant Colonel

Colonel

Brigadier General

Major General

Lieutenant General

General

General of the Air Force

An air force member's pay depends on his or her rank and years of service. Service members with the same rank and years of service are paid equally. Promotion opportunities are based on performance. Enlisted airmen who earn a bachelor's degree can be promoted to commissioned officer positions.

LIFE IN THE AIR FORCE

The US Air Force has a variety of aircraft under its command that operate from bases around the world. The most well-known air force aircraft are fighters, bombers, tankers, and cargo planes. At the end of 2018, the air force had 1,473 fighters and 140 bombers active in its fleet, along with 441 tankers and 278 cargo planes.[6] Each type of aircraft has its own function and role in the air force's operations.

The primary focus of daily life in the US Air Force is living and working on an air base. Air bases are located around the world. Airmen are assigned to a base that needs their skills, but they can change base locations as often as every three years. Each air base is like a small city. It provides its airmen with a variety of amenities and support for themselves and their families, including housing, recreation facilities, shopping, and medical facilities.

Every 20 months or so, an airman can be deployed to an active mission site.[7] In 2017, Staff Sergeant Rhoniesha Seubert was deployed to Al Dhafra Air Base, about 20 miles (32 km) south of Abu Dhabi in the United Arab Emirates. For six months, Seubert worked nights in a dimly lit office, using a computer to coordinate and log the movement of people and weapons in the base's munitions storage area so that the base's bombs and other weapons were safely accounted for at all times. When not working, she exercised or went to the base's Community Activity Center, where she could hang out with friends or watch a movie. Her room, which she shared with a roommate, was tightly crammed with two twin beds and two tall storage chests. There was barely enough space to walk between them. Sometimes she used an iPad to FaceTime with her husband at home or watch Netflix. While the days were tiring, Seubert enjoyed the work. "It makes me feel as if we have a purpose," Seubert said about her accounting for the base's bombs and other munitions. "It's rewarding knowing we're supporting arming aircraft."[8]

KEEPING IN TOUCH

Air force personnel are stationed all over the world. Sometimes they are in very remote locations. Yet no matter where they are stationed, the air force enables its service members to keep in touch with family and friends by phone, email, video chat, and instant messaging. In addition, airmen receive 30 days of paid vacation each year. They can fly for free, which makes it affordable to meet family or friends at home or on vacation.

WELCOME
TO BASIC

TRAINING AT LACKLAND

All air force enlisted recruits attend eight and one-half weeks of basic military training, also known as boot camp, at Lackland Air Force Base in Texas. When recruits arrive for processing week, they fill out forms, have medical and dental exams, receive vaccinations, and have their hair cut. They receive uniforms and running shoes, and they undergo fitness testing.

Once this processing week is complete, the first week of training begins. Over the next seven weeks, recruits spend time in the classroom learning about a variety of topics, from air force history to cyber awareness. In addition, recruits participate in a rigorous physical fitness program designed to improve their strength and endurance. They also learn drill (marching) skills, small arms marksmanship, and first aid. The training days are intense with little free time.

Recruits spend their seventh week of training in the field. They participate in challenging exercises and combat scenarios. Once recruits have completed basic training, they earn the privilege of wearing the blue uniform of the US Air Force, signifying that they have become airmen.

The formal Airman's Coin Ceremony in the eighth week of training honors recruits' hard work and dedication. Each airman receives his or her Airman's Coin. Family and friends are invited to celebrate a recruit's graduation. The following day, graduating recruits march in a parade. The parade ends with a ceremony in which the new airmen recite the US military's oath of enlistment. After completing training, each new airman moves on to one of several jobs in the air force.

 An air force staff sergeant examines a lineup of recruits in basic training.

CHAPTER 4

PILOT

The US Air Force flies some of the most advanced aircraft in the world. From the cockpit, air force pilots control each aircraft and fly missions around the world. These skilled professionals serve as leaders in the air force, training and commanding crews on the ground and in the air.

A US Air Force pilot prepares for takeoff in an F-15E Strike Eagle plane.

FLYING HIGH

Pilots fly a variety of aircraft in the flight operations of the US Air Force. They fly fighters, bombers, transport planes, tankers, and other support aircraft. Some pilots even fly remotely piloted aircraft. Some pilots are combat flyers and fly missions into dangerous situations. They may fly fighter planes and engage in air-to-air combat with enemy planes. They may perform surface attack missions or escort bombers or other aircraft. Some combat pilots fly planes that carry bombs, torpedoes, or

★ Air force pilots fly many different advanced aircraft, including the C-17 Globemaster III, which is used to transport troops and cargo.

missiles. Other combat pilots fly missions to support and protect ground troops.

Air force pilots fly noncombat missions as well. Pilots fly transport aircraft to move troops and supplies to locations around the world. They conduct search-and-rescue missions, reconnaissance missions, and surveillance missions. They perform aerial refueling, filling up other planes with fuel while in the sky. Some pilots fly experimental aircraft to test new technology. Pilots also perform humanitarian missions to deliver supplies to people in need worldwide.

But pilots do more than fly aircraft. "Flying is only about 20 percent of what we do," says Captain Brian Boardman, a 41st Airlift Squadron pilot based at Little Rock Air Force Base in Arkansas.[1] Each week, pilots receive their schedule, which details what they will be doing, including missions, training flights, ground events, and more. If scheduled to fly, the pilot plans the mission with other members of the flight crew. To start, they identify the mission's objectives. They gather information about the weather, determine how much fuel will be needed, and create an emergency plan. Today, air force pilots typically plan missions on a computer and then load the plans, such as flight route and weapons information, onto a data transfer cartridge (DTC). The DTC is loaded into the jet and transfers all of the planning information into the aircraft's computer systems. The crew also creates paper maps and other documents that

REMOTELY PILOTED AIRCRAFT

Remotely piloted aircraft, or unmanned aerial vehicles, are aircraft that fly without anyone on board. A pilot flies the remotely piloted aircraft from a ground station that can be thousands of miles away. He or she uses remote controls to tell the plane what to do. Because the pilot is not on board, the plane can fly on the most dangerous missions. If the plane goes down, the pilot stays safe at the ground station. In the air force, remotely piloted aircraft fly surveillance missions and send pictures and videos to military commanders. Some remotely piloted aircraft carry weapons. They use their cameras to find and identify a target. Then the pilot can remotely launch missiles or drop bombs from the remotely piloted aircraft to attack targets on the ground.

contain all of the mission's important information, such as takeoff and landing calculations, code words, a route overview, and radio frequencies.

After planning is complete, the pilot in charge of the lead aircraft leads a flight briefing meeting a few hours before takeoff. At the briefing, all members of the flight gather and go over the exact details of the mission. They make sure everyone knows what will happen, when it will happen, and how it will happen.

When it is time to fly, pilots check in with the squadron's operations desk and receive a brief on the specific planes they will be flying as well as the latest weather, airspace conditions, and even bird conditions. Then they head out to their aircraft. They perform a walk-around and check for any visible sign of damage or discrepancies in the plane that would compromise the mission. After the pilot and crew confirm that everything is good, they take off on their mission. The mission can be anything from engaging in combat to conducting surveillance, transporting troops and supplies, delivering humanitarian aid, testing new technology, and more.

After the mission is complete, the pilot prepares for the mission debrief. The debrief examines everything that happened that day. It reviews the actions of every flight member and identifies any mistakes that were made. Understanding why a

★ All air force pilots are in the officer ranks.

mistake occurred and how it affected the mission allows the pilot to improve his or her performance for the next mission.

Captain Boardman flies a C-130, which is a military transport aircraft. Even though it's hard work, Boardman says being a pilot is never dull. "There are always opportunities to do something different, and experience something you have never done before," he says. "The days we are able to fly can be considered a break from the office, a chance for us to go out and perform the core job we were all trained to do."[2]

WORK ENVIRONMENT

Air force pilots are assigned to air force bases in the United States or overseas. They spend much of their time on base, preparing for flights often as part of a small team. In the air, pilots may spend many hours in a cramped cockpit, depending on the length of the mission. Weather conditions may vary, which can lead to some rough and bumpy flights.

Captain Jeremy Nolting flies an F-16 fighter jet. He is a pilot with the 79th Fighter Squadron, based at Shaw Air Force Base, South Carolina. One of his favorite parts of the job is the opportunities it gives him to travel. "I get to jet set all around the

MAKING AN IMPACT

The US Air Force has a long history of using its planes and pilots to deliver humanitarian aid to people in need around the world. In 2019, the US Air Force airlifted tons of humanitarian aid to Cucuta, Colombia, a town on the border with Venezuela. The border city was crowded with migrants from Venezuela who were fleeing their country's unrest. Three US Air Force cargo planes took off from Homestead Air Reserve Base in Florida carrying needed supplies to help the migrants. "United States Air Force humanitarian aid missions are the most meaningful missions that we fly," said Captain Susan Jennie, a C-17 pilot with the 6th Airlift Squadron. "The opportunity to fly these kinds of missions was my biggest motivation to train and fly on the C-17. To be able to help out and bring aid to those in need, when needed, is one of the most rewarding opportunities I've been presented in my life and career."[3]

world to cool and exotic places, eat new and interesting foods, meet a myriad of different people," he says.[4]

QUALIFICATIONS AND EDUCATION

Air force pilots are commissioned officers. As such, they are required to have a bachelor's degree from an accredited four-year college or university. They may also come out of the US Air Force Academy, which graduates hundreds of pilot trainees each year. No specific major is required to become a pilot. However, a degree in aviation, aerospace engineering, physics, computer science, or chemistry may be helpful to someone pursuing a pilot career. Candidates with civilian flight training may also have an advantage during the pilot selection process. Aspiring pilots must start pilot training between the ages of 18 and 33. They must also meet certain minimum scores on the Air Force Officer Qualifying Test, which is an aptitude test that is similar to the SAT.

SELECTION BOARD

A recruit who wants to become an air force pilot must submit an application to a selection board. Selection is competitive. A board of senior officers reviews received applications twice a year in February and August. They evaluate applicants on their character, academic achievement, community service, and leadership potential. They review an applicant's test scores and his or her work experience, performance reports, and recommendations. The selection board decides which candidates can train to become air force pilots.

Pilot candidates must meet certain physical standards. They need to pass a variety of physical, psychological, and background tests. Height requirements vary by aircraft. Candidates who are significantly shorter or taller than average may be screened to ensure they can safely operate an aircraft. In addition, air force pilots must have normal color vision and uncorrected vision that is no worse than 20/30 for near vision and 20/70 for distant vision and correctable to 20/20. Candidates with poor depth perception or who have had laser eye surgery may be disqualified from this career. Candidates will also be disqualified for having a history of hay fever, allergies, or asthma after age 12.

TRAINING AND ADVANCEMENT

Pilot candidates who are not already commissioned officers first attend Officer Training School at Maxwell Air Force Base in Montgomery, Alabama. The intense 9.5-week training is designed to challenge potential officers mentally and physically. Candidates participate in daily physical conditioning. They also have a variety of classroom lessons and field exercises to learn navigation, first aid, self-defense, tactical marches to move troops from one location to another when the enemy is near, and base defense.

Next, pilot candidates attend introductory flight training, a course for Reserve Officer Training Corps (ROTC) or Officer Training School graduates who do not have a civilian pilot's license. Flight instructors teach candidates the basics of flying using a small, single-engine plane. Students receive 25 hours of classroom instruction and 17 hours of required flying time, and they must fly solo at least once.

Candidates then attend a yearlong program of specialized undergraduate pilot training (UPT) at one of three air force bases in Mississippi, Texas, or Oklahoma. Candidates spend 10 to 12 hours per day in the classroom, in simulator training, and in the air, as they learn basic flight skills. They learn navigation, aerobatics, formation flying, and how to use aircraft instruments. Then candidates enter one of four advanced training tracks where they learn how to fly a specific type of aircraft, such as fighters or tankers. Some officers may

AIR FORCE ROTC

Some students choose to enter Reserve Officer Training Corps (ROTC) programs at their colleges and universities. ROTC is offered at more than 1,700 colleges and universities in the United States.[5] It prepares college students to become officers in the US military. When students join ROTC, the military pays for their college education. In return, the students commit to serving in the military after they graduate. Along with the air force, other branches of the military also have ROTC programs.

★ Air force pilots have many checklists and procedures to follow to make sure they stay safe while flying.

be chosen to operate remotely piloted aircraft. Upon successful completion of UPT, officers become air force pilots.

For most air force officers, the service commitment is four years of active duty. For pilots, however, the service commitment is longer. Navigators commit to six years of active duty service beginning when they complete training. Pilots commit to ten years of active duty service from the time they complete training.

TOP FIVE QUESTIONS

★ WHAT IS IT LIKE TO WORK AS A PILOT IN THE AIR FORCE?

Air force pilots can be stationed anywhere around the world, sometimes in remote or undesirable locations. When they are not flying planes, they spend a lot of time planning for missions, reviewing completed missions, and training.

★ WHAT SHOULD THOSE WHO ARE INTERESTED IN THIS CAREER STUDY IN SCHOOL?

Students interested in becoming air force pilots should take science classes. They should also study engineering and aviation if possible.

★ WHAT SKILLS ARE BENEFICIAL FOR THOSE INTERESTED IN THIS CAREER?

Pilot candidates should have a knowledge of planes and aviation. A civilian pilot's license is also beneficial. They should also be able to remain calm and work under pressure.

★ HOW ARE AIR FORCE PILOTS PAID?

In the air force, basic pay is based on a service member's rank and years of service. As air force officers advance in rank and years of service, their base pay increases. In addition to base pay, pilots are eligible for bonuses of up to $35,000 per year.[6] The air force offers these bonuses in an attempt to keep experienced pilots in the military.

★ AFTER SERVING IN THE AIR FORCE, WHAT CIVILIAN CAREERS ARE PILOTS QUALIFIED FOR?

After the air force, pilots are qualified to fly commercial airliners, corporate jets, cargo freighters, and more. Some air force pilots choose to become civilian flight instructors to teach others basic flight skills.

CHAPTER 5

TACTICAL AIRCRAFT MAINTENANCE SPECIALIST

The US Air Force uses thousands of aircraft every day. It takes a lot of work to keep the air force's high-tech aircraft in top shape. It's the job of aircraft maintenance specialists to

A tactical aircraft maintenance specialist works on an F-16C Fighting Falcon fighter jet.

make sure that every part of these high-tech aircraft performs to the air force's high standards.

Tactical aircraft maintenance specialists focus their skills on the air force's tactical aircraft, which includes its fighters, strike-fighters, and attack planes. Fighter planes are designed for aerial combat, while attack planes are used to hit ground targets. Strike-fighters can attack ground targets and also engage in air-to-air combat.

MAINTAINING AIRCRAFT

Also known as crew chiefs, tactical aircraft maintenance specialists are generalists who coordinate the tactical aircraft's care and maintenance. They perform general maintenance and call in specialized technicians when necessary. When aircraft are not in the air, tactical aircraft maintenance specialists make sure the planes are maintained and repaired so they are ready to go on a moment's notice. Tactical aircraft maintenance specialists are responsible for inspecting every part of an aircraft to make sure it is ready to safely fly.

Tactical aircraft maintenance specialists inspect aircraft before and after a flight. When an aircraft returns from flight, tactical aircraft maintenance specialists look at the plane's landing gear and engine. They inspect every part of the aircraft, including its structures, systems, and components. In addition, tactical aircraft maintenance specialists review electronic maintenance data. They use that information to identify, diagnose, and resolve any issues with a plane.

After a plane is repaired, tactical aircraft maintenance specialists test repaired components and systems to make sure the aircraft is safe to fly. They log and track all maintenance information so that it can be monitored to identify any patterns in maintenance problems or repairs.

Tactical aircraft maintenance specialists are also responsible for scheduling routine maintenance for all of the air force's aircraft. They make sure all required maintenance is performed on time. They also assist in launching and recovering aircraft. When a plane crashes, tactical aircraft maintenance specialists are part of the crash recovery team.

Staff Sergeant Lee Taylor is a tactical aircraft maintenance specialist with the 15th Aircraft Maintenance Squadron. "As a crew chief, it's my job to perform [aircraft] servicing procedures, as well as aircraft inspections and any other maintenance that needs to be done to keep the aircraft flying," he says.[1] A typical day for Taylor starts by receiving an update from the previous shift's crew. After that, Taylor and other airmen head to the aircraft to perform the day's maintenance tasks.

With many technicians working on a single aircraft, good communication is essential. "It's important that we communicate

MAKING AN IMPACT

Without functioning aircraft, the US Air Force would not be able to fly its missions and protect the United States. Every day, tactical aircraft maintenance specialists work for many hours behind the scenes to ensure that pilots and aircrews can take off on missions around the world at a moment's notice. Without their efforts to keep the air force's fighters, strike-fighters, and attack planes running smoothly, the lives of many airmen would be at risk.

★ Aircraft maintenance specialists may use a log to keep track of their work.

with the specialists to make sure all of our maintenance procedures are performed," Taylor says. Taylor's goal is to complete the specialized training necessary to become a flying crew chief, which would allow him to fly with the aircraft as part of the crew and ensure the plane is operating safely during a mission. It would give him the opportunity to travel more, which he enjoys. "My favorite part of my job is definitely the travel," he says. "I've been afforded a lot of great travel opportunities because of this job." After a busy day overseeing several maintenance projects, Taylor feels a sense of accomplishment. "I have a great sense of pride as a crew chief because I can really see what we do and see the mission happen as a result of what we do," he says.[2]

Tactical aircraft maintenance specialists typically work on air force bases in airplane hangars, repair stations, or airfields. They often need to lift heavy objects, operate power tools, and handle dangerous chemicals. They also climb scaffolds or ladders and are exposed to noise and vibrations, especially during engine testing. To protect themselves from injury, they take precautions such as wearing ear protection while on the job. With specialized training, some tactical aircraft maintenance specialists can work as part of an aircraft's crew, making sure the plane operates safely as it flies on a mission.

QUALIFICATIONS AND EDUCATION

Tactical aircraft maintenance specialists are enlisted airmen and need to have a high school diploma or a GED. Classes in physics and electronics are beneficial. Those entering this career must be between the ages of 17 and 39 at the time they enlist. They must also be able to meet the air force's physical

★ An aircraft maintenance specialist examines the cockpit of a fighter aircraft.

fitness standards. Like many air force jobs, tactical aircraft maintenance specialists must also have normal color vision.

Candidates interested in this job should have mechanical skills and an interest in electronics and repair. A knowledge of aircraft systems, engines, and other components is also beneficial. Candidates also need to qualify for a secret security clearance from the Department of Defense, passing a background check of character and finances. A history of drug and alcohol abuse or a criminal record may prevent a person from entering this career.

Staff Sergeant Beau Blackburn, a crew chief for the 35th Aircraft Maintenance Squadron, knew from a young age that he loved the hands-on nature of maintenance work. "I just wanted to get my hands dirty," Blackburn says. "I enjoyed working on cars before joining the military, so I figured I would see what working on airplanes was all about. I like the idea of discovering something is broken and then doing what I have to in order to fix it."[3]

EARNING CERTIFICATIONS

For many air force careers, earning certifications demonstrates that an airman has mastered certain skills. This can make them more likely to be promoted. For tactical aircraft maintenance specialists, there are many certifications available, including certified aerospace technician, certified production technician, and certified manager.

TRAINING AND ADVANCEMENT

After completing basic training and Airman's Week training, tactical aircraft maintenance candidates attend technical school at Sheppard Air Force Base in Texas. At this school, airmen learn the basics of aircraft maintenance. They learn every part of

AIRMAN'S WEEK

After completing basic military training but before heading to technical training for their specific jobs, airmen attend a special training called Airman's Week. This one-week program is designed to reinforce US Air Force core values and character development for all newly graduated airmen. During the week, new airmen prepare to transition to technical training. They are placed in squadrons with other new airmen who have similar technical specialties.

the plane, including its hydraulic systems, engines, landing gear, fuel systems, and electrical systems.

The length of this training is determined by the type of aircraft the candidate is working on. For example, those who are assigned to work on F-16 fighter jets will go to Holloman Air Force Base in New Mexico for additional training. While the Sheppard training is mostly classroom-based, at Holloman students get four weeks of hands-on training on an operational aircraft. After finishing this training, students earn their skill level 3 (apprentice) rank and are assigned to a duty location to continue on-the-job training.

TOP FIVE QUESTIONS

★ WHAT IS IT LIKE TO WORK AS A TACTICAL AIRCRAFT MAINTENANCE SPECIALIST?

Tactical aircraft maintenance specialists make sure the air force's aircraft are operating smoothly and safely. They make sure every single component of an aircraft is functioning properly and the aircraft will be able to complete its mission.

★ WHAT SHOULD THOSE WHO ARE INTERESTED IN THIS CAREER STUDY IN SCHOOL?

Students who are interested in becoming tactical aircraft maintenance specialists should take classes in computer science, physics, engineering, electronics, and maintenance and repair.

★ WHAT SKILLS ARE BENEFICIAL FOR THOSE INTERESTED IN THIS CAREER?

Tactical aircraft maintenance specialists should have mechanical and electronics skills to help them repair and maintain the air force's high-tech aircraft. They should also be very detail-oriented, as the tiniest problem with an aircraft component may endanger the aircraft's mission and its crew.

★ HOW ARE TACTICAL AIRCRAFT MAINTENANCE SPECIALISTS PAID?

In the air force, basic pay is based on a service member's rank and years of service. For example, a tactical aircraft maintenance specialist who has more than eight years of service can earn $3,945 per month.[4] In addition to base pay, tactical aircraft maintenance specialists, like all enlisted airmen, receive other compensation in the form of housing, food, and uniform allowances.

★ AFTER THE AIR FORCE, WHAT CIVILIAN CAREERS ARE TACTICAL AIRCRAFT MAINTENANCE SPECIALISTS QUALIFIED FOR?

Tactical aircraft maintenance specialists can work as aircraft and avionics equipment mechanics or technicians. They are often employed by airlines, specialized repair and maintenance companies, aerospace manufacturing companies, and the federal government.

CHAPTER 6

SURVIVAL, EVASION, RESISTANCE, AND ESCAPE SPECIALIST

How does an airman know what to do when downed behind enemy lines or stranded in a hostile environment? Survival, evasion, resistance, and escape (SERE) specialists teach air force personnel how to handle and react to any

SERE specialists work to train other airmen to survive in any environment.

environment, in any conditions. With the training provided by SERE specialists, airmen have the skills and knowledge needed to survive if their aircraft goes down.

TRAINING OTHERS TO SURVIVE

The air force's SERE specialists are part of the Air Force Special Ops. Unlike other special operations personnel such as combat controllers or pararescuemen, SERE specialists are not involved in direct combat situations. Instead, they provide the training, skills, and knowledge that other airmen need to survive any situation. "We're not a combat career field. We're not down range shooting people. We're not kicking doors down. We're instructors," says Justin Samaniego, a SERE specialist orientation course unit training manager.[1]

SERE specialists are experts on how to survive anywhere on Earth. Their motto is "return with honor," with the goal of expending all possible effort to successfully complete every part of a mission. SERE specialists' goal is to teach airmen the skills they will need to return home from missions safely. SERE specialists teach airmen how to survive in captivity or when isolated in any environment, such as arctic areas, deserts, jungles, mountain regions, and even on the open seas. They teach airmen how to find food in nature, find or build a shelter, and start a fire. They also teach them how to avoid pursuit and capture by the enemy and how to respond if captured.

Sometimes SERE specialists are deployed to combat locations. Although they do not participate directly in combat operations, they serve as experts to help plan recovery and rescue operations. For example, a SERE specialist can advise a pararescue team about how long a downed pilot can survive in specific conditions as they make their rescue plan. SERE specialists can also train the rescuers in using available equipment. "It is our job as SERE specialists to ensure the

★ A SERE specialist shows trainees how to use a compass for land navigation.

tactics, techniques and procedures we teach gives anyone who goes through our course the necessary skills and confidence needed to return with honor, regardless of the circumstances of their isolation," said Senior Master Sergeant Brian Kemmer.[2]

Master Sergeant Bob Miner is a SERE specialist who trains members of the West Virginia Air National Guard. Miner's students include all members of a flight crew. To prepare the crew for any emergency, Miner teaches the most up-to-date training in local area survival, combat survival, conduct after capture, water survival, and emergency parachute training. Miner also helps the unit with preparing reports, evasive plans of action, and personnel recovery kits. Miner helps out with survival training for other units that do not have a dedicated SERE specialist, too. It's a job that Miner is proud of doing. He not only gets to travel the country but also knows he's making a difference. If something happens to one of his students when they are on a mission, Miner's training may be the reason they make it home alive.

WORK ENVIRONMENT

SERE specialists work anywhere in the world. They can be assigned to train squadrons in the United States or overseas. Most of the time SERE specialists work outdoors to teach

MAKING AN IMPACT

In 1986, air force pilot Dale Storr took a SERE survival course. About five years later, a SERE specialist visited Storr's base and taught a refresher class on how to survive capture. Months after the refresher class, during Operation Desert Storm, Storr was shot down and captured by an enemy. Storr remembered his SERE training and played the role of a good prisoner. When asked to draw the layout of his base in Saudi Arabia, Storr pretended to comply but drew another base in Oklahoma instead, one which he knew extremely well. Throughout his 33 days of captivity, Storr did not reveal any information that could possibly harm American lives. "If I didn't have any survival training, I probably would have still survived," says Storr. "But I'd be a mental wreck today. I would have blabbed all this information and not had any way to resist. I can't imagine what that would feel like." Storr is one of many airmen who have survived dangerous situations using skills they learned in SERE training.[3]

survival skills. They often endure the harshest conditions in all climates, from the Arctic to deserts to the tropics, as they train others.

Ben Domian is an air force reserve SERE specialist and has taught survival skills to aircrew members for more than eight years. As a third-generation air force member, he has traveled around the world to train and teach survival in the most grueling conditions. He says:

> *I love my job, and I tell people that if they weren't paying me, I would pay them because I absolutely love it. We jump out into the middle of the ocean, inflate boats, and drive them to shore. I have had some amazing experiences, like conditions of negative 46 degrees [-43°C] in Alaska, building igloos, and doing survival training where your nose hairs freeze together. Then the deserts of Africa, 20-foot [6 m] seas in the Pacific Ocean, the jungles of Korea, and now the jungles of Fiji . . . so I have done survival in all four corners of the world.*[4]

EDUCATION AND TRAINING

SERE specialists are enlisted airmen. As such, they are required to have a high school diploma. Because the job of a SERE specialist is physically demanding, those interested in this career must meet certain physical standards. They must be able to successfully complete the SERE physical ability and stamina test (PAST). Candidates must complete a 200-meter surface swim in under ten minutes, run 1.5 miles (2.4 km) in less than 11 minutes, perform certain calisthenic exercises, and complete a minimum of eight pull-ups in one minute, 48 sit-ups in two minutes, and 48 push-ups in two minutes. In addition, SERE candidates must be able to complete a four-mile (6.4 km) march carrying a 60- to 65-pound (27 to 29 kg) pack in 60 minutes or less. Once a candidate successfully completes basic training and meets these requirements, he or she can begin SERE training.[5] SERE candidates must be able to speak clearly and have normal night and color vision. They also need to be able to lift 70 pounds (32 kg) overhead and have no allergies to pollens or grasses.[6]

ARMED SERVICES VOCATIONAL APTITUDE BATTERY

When they enlist, all military recruits, regardless of service branch, take the Armed Services Vocational Aptitude Battery (ASVAB). The ASVAB is a multiple-choice test that measures a recruit's strengths, weaknesses, and potential for future success in the military. Developed by the Department of Defense, the ASVAB tests recruits' knowledge in several subjects such as science, arithmetic reasoning, word knowledge, and electronics and mechanical comprehension. The military uses the test to determine if a recruit has the mental aptitude to enlist and succeed in the military. It can also help determine which military careers are the best fit for the recruit. Some air force careers require recruits to meet a minimum ASVAB score.

After basic training, SERE candidates stay at Lackland Air Force Base in Texas for the SERE Specialist Screening course, which lasts for 19 days. The course tests their physical and psychological limits and evaluates their speaking abilities, leadership skills, dedication, and abilities to complete tasks and follow instructions. Only the top candidates are chosen to move on to formal SERE training.

After they pass the screening, SERE candidates travel to Fairchild Air Force Base in Washington for SERE Specialist Training (SST). SST is a five- to six-month intense survival training course. Candidates go through the course as students so that they can become survival experts and teach others. Candidates learn to survive in various environments, including the mountains, tropics, deserts, and oceans. They train in arctic conditions at Eielson

AIR FORCE SPECIALTY CODES

In the air force, enlisted jobs are known by their Air Force Specialty Codes (AFSCs). AFSCs are divided into seven categories: operations, maintenance and logistics, support, medical and dental, legal and chaplain, finance and contracting, and special investigations. Within each category, jobs are further classified into career fields. A career field may include a single AFSC or multiple AFSCs. For example, the AFSC for a SERE specialist might be 1T071, in which 1 represents the career group (operations), T represents the career field (aircrew protection), 0 represents the career field subgroup, 7 represents a skill level (craftsman), and 1 represents a specialty within the career field.

Air Force Base in Alaska and learn water survival skills in Pensacola, Florida.

About three months of SST is in the field, where training and instruction occur in all major climates. The training often takes place in remote locations, and candidates experience all potential weather conditions. They learn wilderness first aid, hand-to-hand-combat, and rough land evacuation tactics. Candidates search for water in the desert. They forage for berries in the forests and mountains. They even practice freeing themselves from a helicopter upside down in a pool. They practice parachute landings in a hangar and learn how to escape as a hostage by crawling through drainage pipes and scaling walls.

"There was one week [during training] where I probably only slept eight hours total," says SERE specialist Peter Ryan. On a training expedition in the Arctic, the temperature never rose above single digits. Ryan struggled on tired legs and chopped wood in snow up to his waist. Every night, he had to dig a new shelter, use the wood to make a fire, and tackle a long list of tasks given to him by his instructor. His reward for finishing it all was simply a few hours of sleep before the next day. "We're the guinea pigs," Ryan says. "I can't teach you what to do as an isolated person if I don't know what it's like myself."[7]

★ SERE specialists are trained to parachute in different environments.

After finishing SST, SERE candidates go to Army Airborne School at Fort Benning, Georgia. There, they practice parachuting skills. After graduating from airborne school, SERE specialists begin on-the-job training to become instructors for the combat SERE course. They start by teaching skills that they have mastered. As they gain experience, they teach more and more skills, until they are teaching all parts of the course. Experienced SERE specialists may also be responsible for planning the training by creating lesson plans and arranging locations for field exercises. After three years as a field instructor, a SERE specialist can qualify to participate in personnel recovery missions.

TOP FIVE QUESTIONS

★ WHAT IS IT LIKE TO WORK AS A SERE SPECIALIST?

SERE specialists must be prepared for any climate, condition, or situation. They become experts in survival by exposing themselves to extreme conditions and situations so that they can better teach survival skills to other air force personnel. Much of their time is spent outdoors, practicing their own skills and training others.

★ WHAT SHOULD PEOPLE WHO ARE INTERESTED IN THIS CAREER STUDY IN SCHOOL?

SERE candidates should take as many physical education classes as possible in high school. They should also take English, mathematics, and basic science courses to prepare for the Armed Services Vocational Aptitude Battery (ASVAB) exam, which they will take upon enlistment.

★ WHAT SKILLS ARE BENEFICIAL FOR THOSE INTERESTED IN THIS CAREER?

SERE candidates should be in top physical condition. They should be able to communicate clearly and effectively. They should also be able to remain calm and work under pressure.

★ HOW ARE SERE SPECIALISTS PAID?

In the air force, basic pay is based on a service member's rank and years of service. SERE specialists, like all enlisted airmen, receive additional compensation in the form of housing, food, and uniform allowances.

★ AFTER SERVING IN THE AIR FORCE, WHAT CIVILIAN CAREERS ARE SERE SPECIALISTS QUALIFIED FOR?

SERE specialists are typically valued for their leadership skills, self-determination, and organization. After the air force, they can use these skills as outdoor guides, as teachers, or in a variety of other careers.

CHAPTER 7

WEATHER OFFICER

Weather can have a significant impact on aircraft performance. Low clouds, fog, and rain can affect visibility. Thunderstorms and lightning can disrupt planned missions and make it more difficult for pilots to fly. For those reasons, accurate weather forecasting is essential to ensuring the air force's missions are successful and safe.

Air force weather officers send weather data to other military personnel to inform missions.

PREDICTING THE WEATHER

Air force weather officers perform, manage, and direct weather operations that affect the activities of all US military forces. They constantly monitor weather in areas where the military operates. They predict weather patterns and prepare weather forecasts. Then they prepare detailed briefings to communicate weather information to commanders and pilots before every mission. "We can give commanders an advanced notification on all weather conditions in their area of operation," says Staff Sergeant Donald

WHAT IS METEOROLOGY?

Meteorology is the study of Earth's atmosphere and its effect on weather. The atmosphere is the gaseous layer that surrounds Earth. The atmosphere extends about 65 to 75 miles (105 to 121 km) above Earth's surface.[2] Meteorology is a branch of atmospheric science, the study of the atmosphere. Meteorologists use scientific principles to observe, explain, and forecast weather.

Johnson, a battlefield weather forecaster with the 18th Weather Squadron.[1]

Weather officers collect, record, and analyze weather data using the latest technology. They use meteorological sensors to measure and evaluate atmospheric conditions. They monitor space weather such as solar storms which can trigger explosions called solar flares that can damage satellites and disrupt power grids. They analyze atmospheric and space data using satellite and radar imagery, computer-generated graphics, and weather communication equipment. Using this data, they prepare forecasts of atmospheric and space weather conditions. When necessary, weather officers issue warnings and advisories to let commanders know if the weather could impact a mission.

In some cases, air force weather officers are attached to army units in the field. Army commanders rely on the accuracy of the weather officers' forecasts in order to effectively deploy assets such as helicopters or remotely piloted aircraft. "The SWO [staff weather officer] tells me what our limits are," says

★ Weather officers work together to gather information such as snowfall measurements.

Second Lieutenant Andrea Nevistic, who is part of the First Brigade Combat Team, 82nd Airborne Division. "Everything from air elements to ground elements are impacted by weather."[3] For weather officers, accurate forecasting is essential. "If we aren't accurate with our forecast, we lose credibility," Johnson says. "Our forecast can also give an idea as to which is the best route to use as well as which route our enemies are likely to use."[4]

Weather officers also develop and coordinate important weather studies and research. First Lieutenant Daniel

Katuzienski is a master's student with the Air Force Institute of Technology. He is researching and developing radar forecasting techniques to better predict and mitigate the impact of lightning on air force space missions. "As you can imagine, lightning and rockets don't necessarily mix. My project focused on optimizing Doppler weather radar parameters to improve radar lightning forecast methods," Katuzienski says.[5]

WORK ENVIRONMENT

Weather officers work on air force bases in the United States and overseas. They primarily work in offices using computers and other technologies to gather weather data and produce forecasts. But sometimes weather officers are deployed with a military unit in the field to provide weather forecasts.

Captain Audra Goldfuss is a weather officer assigned to the Seventh Air Force in Osan, South Korea. Her typical day begins at 5:30 a.m. when she reports to work. She is briefed on the previous 12 hours of weather in her base's area of responsibility, which includes the Korean Peninsula, Japan, Hawaii, Guam, and parts of China. She reports to a briefing room and gives an update to the general in charge of the base. While the weather is not top secret, the briefing requires a high security clearance because of how weather impacts the aircraft the base flies. In addition to analyzing weather data and preparing forecasts,

★ Weather officers may work outside or in an office.

Goldfuss takes online military education courses. Along with all base personnel, she participates in regular combat exercises. "We practice as if bombs are dropping, guns are firing. . . . My job during these times is to give weather for specific locations, help out with personnel recovery missions, and keep people informed of weather all over the [area]," she says.[6]

EDUCATION AND TRAINING

Weather officers are commissioned officers. They are required to have a bachelor's degree in meteorology, atmospheric science, or a related field from an accredited four-year college or university. Weather officers must be between the ages of 18 and 39 when they join the air force. They must also be US citizens and must be willing to serve a minimum of four years of active duty service and four years of reserve service.

Like other air force officers, weather officers start their training in Officer Training School, as members of ROTC programs, or at the US Air Force Academy. Then they take the Weather Officer Course at Keesler Air Force Base in Mississippi. Upon completing this training, a weather officer receives his or her first assignment at an operational weather squadron at an air force base. The officers receive additional on-the-job instruction. As they gain experience, weather officers can earn certifications and attend advanced training that will help them advance in rank and receive promotions. Weather officers may also be assigned short-term deployments to support field military operations.

MAKING AN IMPACT

The work done by air force weather officers can save lives, whether on the battlefield or during a natural disaster. When a hurricane approaches, most people try to get out of its way. Air Force Major Ashley Lundry, an aerial reconnaissance weather officer, does the opposite. She flies into the storm. Lundry is a member of the Air Force's 53rd Weather Reconnaissance Squadron, also known as the Hurricane Hunters. She flies reconnaissance missions directly into severe tropical weather during hurricane season so that she can gather data for the National Hurricane Center. The National Hurricane Center uses this valuable data to improve their forecasts and issue hurricane warnings, which can save lives on the ground and minimize property damage.

TOP FIVE QUESTIONS

★ WHAT IS IT LIKE TO WORK AS A WEATHER OFFICER?

Weather officers use the latest technology to collect and analyze weather data. They use this information to prepare weather forecasts that are essential for planning missions.

★ WHAT SHOULD THOSE WHO ARE INTERESTED IN THIS CAREER STUDY IN SCHOOL?

Weather officers are required to have a bachelor's degree in meteorology or atmospheric science. Students who are interested in this career should study advanced mathematics and physics, along with courses in meteorology and atmospheric science. Classes in computer science and programming are also beneficial.

★ WHAT SKILLS ARE BENEFICIAL FOR THOSE INTERESTED IN THIS CAREER?

Weather officers should have good analytical and critical-thinking skills to work with computer models and data to prepare weather forecasts and determine the most likely weather outcomes. Communication skills are also important as weather officers must be able to write and speak clearly to communicate their forecasts to commanders. Superior math skills are also essential as weather officers use the principles of calculus, statistics, and other advanced math to develop weather forecasting models.

★ HOW ARE AIR FORCE WEATHER OFFICERS PAID?

In addition to basic pay, weather officers, like all other air force officers, receive additional compensation in the form of housing, food, and uniform allowances.

★ AFTER THE AIR FORCE, WHAT CIVILIAN CAREERS ARE WEATHER OFFICERS QUALIFIED FOR?

Air force weather officers can go on to work as weather forecasters for the National Weather Service in weather stations across the country, as broadcast forecasters for television stations, or as forecasters for other private businesses. They may also become involved in weather-related research for a college or university.

CHAPTER 8

CYBER SURETY SPECIALIST

The US Air Force relies on a variety of digital technologies, from sophisticated computer systems and networks to databases and voice systems. While these technologies make many tasks easier and more efficient, they can present security risks. Cyber surety specialists are the air force's information

Cyber surety specialists work to keep the US Air Force secure against cyberattacks.

technology (IT) specialists. They ensure the security of the air force's computer systems, networks, databases, and digital information.

ON THE JOB

The job of a cyber surety specialist in the air force is very similar to that of an IT specialist in a civilian company. The main goal of cyber surety specialists is to ensure the air force's computer systems, networks, and other digital assets remain safe and secure. To do this, they monitor and maintain systems.

They review IT policies and procedures to ensure all systems are adequately protected from unauthorized activity. Cyber surety specialists also conduct risk and vulnerability assessments of the air force's computers and systems. They make sure that air force policies are in compliance with all legal and regulatory requirements and that all policies are enforced.

One critical part of the job is identifying security weaknesses in IT systems. According to Major General Patrick Higby, the US Air Force's director of cyberspace strategy and policy, cyber surety specialists should think like hackers to identify any weaknesses in the air force's cyber defenses. "That causes you to say, 'Oh, wait a minute, I need to pay attention to this [computer] setting,' or 'I need to make sure that these ports are actually closed,' because that's the way the adversary is going

WHAT IS A HACKER?

A hacker is any person who uses computers to explore computer networks. The most well-known type of hacker is called a black hat. Black hats use their computer skills to sabotage computer systems, spread computer viruses, and steal valuable information that they can sell. Other hackers use their computer skills to defend others instead. Organizations invite these hackers, called white hats, to hack their computer networks in order to find security weaknesses so they can be fixed. In hacking, however, everything is not black and white. Some hackers fall in the middle. They break into computer networks without permission, which is a serious crime. However, they do it for fame or the challenge, not for money or sabotage.

to come in to try to attack us," Higby says.[1] Once a weakness is identified, cyber surety specialists come up with a solution to eliminate it. They investigate any security breaches and determine how to improve the branch's cybersecurity.

Staff Sergeant Brandon Whittaker is a cyber surety technician with the 36th Communication Squadron. Stationed at Andersen Air Force Base in the US territory of Guam, Whittaker's team keeps an eye out for potential communication threats. They make sure the base's computer systems comply with Department of Defense and air force computer security policies. Whittaker's team regularly performs operational readiness exercises. The airmen practice defending against phishing attacks, hackers trying to access the network, and the unauthorized release of classified information. During exercises, the airmen make sure authorized users are using good security practices and not causing any weaknesses that unauthorized users could use to gain access to the base's systems. "One of the rewards in my job is being able to see directly how we affect the mission," Whittaker said. "If we don't do our job correctly, there would be countless consequences. We would lose our ability to protect communications."[2]

Every month, Whittaker's team shares cyber awareness bulletins and conducts training to keep base staff up-to-date on the latest best practices in cybersecurity. "The hardest thing

★ Cyber surety specialists sometimes work together to solve problems.

about my job is making sure the information that I give to people is correct," Whittaker said. "Technology is always growing, so as it grows, we will grow. Asking questions, networking and researching helps me stay ahead."[3]

Cyber surety specialists can work at air force bases around the world. Many are assigned to the 24th Air Force, which is the air force's command responsible for cyberspace operations and defense. Others are assigned to other units. Regardless of their assignments, cyber surety specialists typically work in clean, air-conditioned offices, computer rooms, and labs with electronic equipment. They sometimes work many hours a day, especially when a security breach has been identified.

EDUCATION AND TRAINING

Cyber surety specialists are enlisted airmen and need to have a high school diploma or a GED. People entering this career must be between the ages of 17 and 39 at the time they enlist. They must also be able to meet the air force's physical fitness standards. In addition, having some knowledge of computer science is beneficial. Because cyber surety specialists have access to sensitive data and information, candidates must obtain top-secret security clearance from the Department of Defense.

MAKING AN IMPACT

As more and more digital assets are connected online and through networks, a security weakness in one small part of the connected system can put the entire system at risk. In addition, software is increasingly vulnerable to cyberattack. As the amount of software that the US Air Force uses on its systems increases, the number of security vulnerabilities also increases. Because the air force is tasked with defending the United States, protecting its cyber systems is critical. Cyber surety specialists are on the front lines of defending the air force's cyber systems. Their work is essential to making sure the country's space and air programs operate securely. They make sure that access to these systems and information does not fall into the hands of unauthorized users.

To obtain this clearance, candidates go through a background check that looks into their finances and personal history.

After completing basic training and Airman's Week, candidates report to technical school at Keesler Air Force Base in Mississippi. For about two months, airmen learn about the air force's computer systems, networks, databases, cybersecurity, and other related topics. They also learn how to solve problems with the air force's computer systems and how to address potential breaches of the branch's computers and networks.

After technical school, airmen are given a permanent duty assignment at an air force base. There, they continue learning on the job. During this phase, cyber surety specialists learn how to maintain and secure the specific systems at their assignment. After several months of on-the-job training under the supervision of a more experienced cyber surety professional, cybersecurity specialists are ready to operate on their own.

The US Air Force prides itself on the training that it provides its airmen, including those in cyber surety. "We're going to give you all the training you need," Higby says. "They're not going to be masters by the time they graduate [from technical training], but they're going to have enough skills to be successful in learning new things at that first base we send them to. There are not a lot of other companies that invest that much time."[4]

TOP FIVE QUESTIONS

★ WHAT IS IT LIKE TO WORK AS A CYBER SURETY SPECIALIST?

Cyber surety specialists are the air force's information technology and cybersecurity experts. They ensure that all air force computers, networks, and systems are secure, and they make sure all air force personnel are following cybersecurity procedures. When a breach or weakness in security is discovered, cyber surety specialists come up with ways to strengthen cyber defenses.

★ WHAT SHOULD THOSE WHO ARE INTERESTED IN THIS CAREER STUDY IN SCHOOL?

Students interested in becoming cyber surety specialists should take classes in math, computer programming, and computer and network technology.

★ WHAT SKILLS ARE BENEFICIAL FOR THOSE INTERESTED IN THIS CAREER?

Cyber surety specialists spend a lot of time working with sophisticated electronic equipment and computer systems. They must have a good understanding of computers, satellite systems, and other high-tech equipment. They should be able to communicate clearly, both in writing and verbally. In addition, cyber surety specialists should be able to think creatively and problem solve to identify and fix weaknesses in cybersecurity.

★ HOW ARE CYBER SURETY SPECIALISTS PAID?

In the air force, basic pay is based on a service member's rank and years of service. For example, a cyber surety specialist at the rank of airman first class who has more than two years of service earns approximately $2,100 each month in base pay.[5] As airmen advance in rank and years of service, their pay increases. Cyber surety specialists, like all enlisted airmen, receive additional compensation in the form of allowances for housing, food, and other living expenses.

★ AFTER THE AIR FORCE, WHAT CIVILIAN CAREERS ARE CYBER SURETY SPECIALISTS QUALIFIED FOR?

Cyber surety specialists are qualified to work in a variety of information technology careers. They may work for private companies, government agencies, or nonprofit organizations.

CHAPTER 9

PARALEGAL SPECIALIST

Like all organizations, the US Air Force and its personnel need the help of legal counsel from time to time. Air force legal departments provide a full range of legal services to the US Air Force and its personnel. Air force legal departments have an extremely varied practice. They provide expertise in military justice, environmental law, labor law, international law, cyber law, and more. Paralegal specialists work under the supervision

Paralegal specialists help other air force personnel navigate a wide range of legal services.

of air force attorneys, called judge advocate generals (JAGs), to ensure that the air force and its personnel have access to the legal services they need.

WORKING ON THE CASE

Like paralegals in the civilian world, air force paralegal specialists assist attorneys in preparing cases. In their job, paralegal specialists have a wide variety of legal duties. They assist JAGs in preparing for court-martial and other cases by researching case law, interviewing witnesses and victims, preparing claims, assessing legal filings, and drafting legal opinions and documents.

When an air force member is seeking legal help, the first person he or she talks to in the JAG office is often a paralegal. The paralegal interviews the service member to assess if he or she is eligible for legal services from the JAG office. Once the service member is accepted as a client, the paralegal consults with him or her to gather facts and background information for

MILITARY COURT-MARTIAL

Every member of the military, including service members in the US Air Force, must abide by the Uniform Code of Military Justice (UCMJ). The UCMJ is a federal law that defines the military justice system and military criminal offenses. Criminal offenses listed in the UCMJ are prosecuted and punished through military courts-martial.

Under military law, a military court-martial is the most severe sanction that a military member can receive. A court-martial conviction is the same as a federal conviction in civilian court. Depending on the offense, a court-martial conviction can result in jail time, a dishonorable discharge from the military, fines, or a reduction in rank. Military courts-martial come in three forms: summary, special, and general. The type of court-martial depends on the severity of the offense and the rank of the person on trial. The most severe offenses, such as murder, rape, and robbery, are usually tried in a general court-martial. A general court-martial is presided over by a military judge. It includes the accused as well as attorneys for the prosecution and defense. A panel of at least five service members who are at least the same rank as the accused acts as a jury in the proceeding.

the case. The paralegal prepares documents such as witness lists or briefs for court proceedings.

Paralegals also answer legal questions and prepare a variety of other legal documents for airmen and their families. They prepare wills, documents needed to sell homes and other assets, and powers of attorney (which appoint someone to manage a person's affairs if they are unable to do so) for service members and their families. Sometimes an air force paralegal will also be a notary public, a person appointed to verify the authenticity of important documents.

Senior Master Sergeant Sylvetris Hlongwane is an air force paralegal manager assigned to the 11th Air Force at Joint Base Elmendorf-Richardson in Alaska. She advises airmen about tax law and how deployments affect taxes. She briefs them on issues such as the consequences of driving under the influence or other misconduct to help keep them out of trouble. Other times, Hlongwane addresses claims, such as when an airman's personal property is damaged or destroyed.

Hlongwane is also involved in questions of operational law, which come up frequently in Alaska. "Operational law impacts how operations are conducted," she explains. One reason it comes up in Alaska is because the air force works with nearby Canadian military forces there. "Those kinds of partnerships determine, for example, how and where we fly,"

Hlongwane says.[1] There are procedures that must be followed and documentation that must be prepared when the 11th Air Force launches US fighter aircraft to intercept Russian military aircraft as they enter the Alaska Air Defense Identification Zone. Therefore, Hlongwane and other paralegals at the base research issues involved with operational, international, and military law. "There's always a representative from the judge advocate's office in the Air Operations Center to advise commanders with that, and we paralegals are researching for them while they're doing so," Hlongwane says.[2]

Hlongwane says her job as a paralegal is never boring. "There's always something new to learn. There's no sitting back and saying 'I've been doing this for five years, there's nothing more I can learn about legal.' I can have 15 drug cases and there's something different to learn in all 15," she says.[3]

WORK ENVIRONMENT

Air force paralegals can be stationed at US Air Force bases in the United States or overseas. Typically, paralegals work in an office setting. Occasionally, they may travel to various sites to gather information, conduct interviews, or review documents. They also spend time in courtrooms, accompanying JAG attorneys to trials and other court proceedings.

★ Air force paralegals sometimes help people prepare various legal documents.

Often, air force paralegals work on teams with other paralegals, JAG attorneys, and other legal support staff. While they typically work full time, paralegal specialists are sometimes required to put in extra hours to meet deadlines.

QUALIFICATIONS AND EDUCATION

Paralegal specialists are enlisted airmen, so they need to have a high school diploma or a GED. People entering this career must be between the ages of 17 and 39 at the time they enlist. They must also be able to meet the air force's physical fitness standards. Candidates should also be able to type well, at a minimum rate of 25 words per minute.

Although no special security clearances are required to become a paralegal specialist, candidates must not have any previous criminal convictions. Exceptions can be made for minor traffic violations and similar issues. Candidates are also expected to pass a screening of their financial background.

Paralegal specialists must be able to work well with others. They should also be comfortable with following orders from superior officers. They should be familiar with using computers for legal research and preparing documents. They also need to be able to communicate clearly and effectively, both verbally and in writing. Paralegal specialists are expected to follow principles of law such as client confidentiality.

TRAINING AND ADVANCEMENT

After completing basic training and Airman's Week training, paralegal specialist candidates spend 35 days in technical training at Maxwell Air Force Base in Alabama. In this training, they learn the protocols of conducting legal proceedings in military court. They review civil law procedures and legal ethics. They study legal claims and military justice. Candidates also learn how to conduct legal research and prepare legal documents according to air force and Department of Defense regulations. They learn military courtroom procedures, which

MAKING AN IMPACT

Senior Master Sergeant Sylvetris Hlongwane, an air force paralegal manager, believes that one of the most important parts of her job is helping crime victims seek justice. It can be extremely challenging. She explains that it can be tough to remain composed when interviewing an emotional victim or witness. "If someone's been assaulted, a part of me just wants to hug them, but that's not what they need. I have to be stoic to an extent and stay strong with them," she explains. Yet despite the challenges, she believes that her work is making a difference for her clients and others. "The most rewarding thing? Justice. [Nothing] is as rewarding as seeing someone who was harmed receive justice," she says.[4]

★ The air force's career paths can prepare people for jobs after their military service or for a long career in the air force.

may be different than those in civilian courts, along with pre- and post-trial administration procedures. Once candidates successfully complete the training, they are officially designated as air force paralegals.

Throughout their careers, paralegal specialists may attend several specialized courses that include training in operational law, environmental law, contract law, will preparation, federal income tax assistance, and victim and witness assistance. Upon successful completion of these advanced courses, paralegals are eligible to earn college credit. Like people in other air force career fields, paralegal specialists who complete certain training requirements can advance to higher skill levels. As they advance, paralegal specialists take on more supervisory and administrative roles.

Over more than 20 years in the air force, Senior Master Sergeant Donte Anderson advanced from working as a beginning paralegal specialist to become the law center superintendent for the Kadena Law Center at the United States' Kadena Air Force Base in Japan. In this role, he advised the staff judge advocate on all matters involving enlisted airmen. He supervised a team of paralegals that serviced the air force's largest combat wing. "I provided guidance and advice on a myriad of issues to the general counsel in support of 25,000 members who helped maintain over $6 billion dollars in aircraft and equipment. I was responsible for supporting a staff of 31, made up of attorneys, paralegals, and US and Japanese civilians. . . . As a recruiter for the paralegal profession, I vetted members for possible inclusion. Finally, I provided mentorship and guidance to junior attorneys and paralegals," he says.[5]

A HIGH-FLYING CAREER

For those who are able to follow military rules, obey orders, and meet certain physical and fitness standards, a career in the US Air Force can be a great choice. Air force personnel travel the country and the world. They receive thorough training on high-tech equipment that prepares them for a career in the air force and future civilian careers. Because there are so many different types of careers in the air force, those interested in joining this branch of the US military can find the perfect fit to match their skills and interests.

TOP FIVE QUESTIONS

★ WHAT IS IT LIKE TO WORK AS AN AIR FORCE PARALEGAL SPECIALIST?

Paralegal specialists assist air force attorneys with all types of legal matters, from military operational issues to legal advice for airmen and their families. The job is very similar to a paralegal job in a civilian law firm.

★ WHAT SHOULD THOSE WHO ARE INTERESTED IN THIS CAREER STUDY IN SCHOOL?

Students interested in becoming air force paralegals should take courses that improve their communication skills and allow them to practice writing and researching. They should also take computer classes so that they are familiar with online research and have strong typing skills.

★ WHAT SKILLS ARE BENEFICIAL FOR THOSE INTERESTED IN THIS CAREER?

Paralegal specialists should have excellent communication skills. They should be skilled researchers, able to look up laws or other information quickly. They should also be comfortable working with a team of legal professionals.

★ HOW ARE AIR FORCE PARALEGAL SPECIALISTS PAID?

In addition to basic pay, paralegal specialists, like all enlisted airmen, receive other compensation in the form of housing, food, and uniform allowances.

★ AFTER THE AIR FORCE, WHAT CIVILIAN CAREERS ARE PARALEGAL SPECIALISTS QUALIFIED FOR?

Air force paralegal specialists are qualified to work as paralegals or legal assistants in private law firms, public defenders' offices, or district attorneys' offices. Some states may require civilian paralegals to obtain additional licensing.

ESSENTIAL ★★ FACTS

US AIR FORCE HISTORY

- ★ 1907: The US Army Signal Corps creates a small aeronautical division.
- ★ 1908: The Signal Corps tests the army's first airplane.
- ★ 1914: The US Congress creates an Aviation Section within the Army Signal Corps.
- ★ 1917: The United States enters World War I and its aeronautical division expands significantly, becoming the US Army Air Service.
- ★ 1941: Japan attacks the Pearl Harbor naval base in Hawaii. The United States enters World War II and the US Army Air Forces play a significant role in the war.
- ★ 1945: US military planes drop atomic bombs on the Japanese cities of Hiroshima and Nagasaki.
- ★ 1947: The US Air Force becomes a separate branch of the US military.
- ★ 1982: The US Air Force Space Command is formed.
- ★ 1991: The US Air Force shows overwhelming power in Operation Desert Storm, and the Gulf War ends in a few months.
- ★ 2000s: The US Air Force continues to produce innovative and sophisticated technology, including remotely piloted aircraft for the country's defense.
- ★ 2019: The Air Force Space Command becomes the US Space Force, a separate branch of the military.

US AIR FORCE ORGANIZATION

The US Air Force is one of six branches of the US military. It is led by the Department of the Air Force, which is part of the US Department of Defense. Air force personnel are divided into 11 Major Commands, which are further divided into numbered air forces and wings. Most wings are subdivided into smaller units called groups. Within each group, personnel are assigned to a squadron. Each squadron has a specialized function, such as a fighter squadron or a training squadron. Each squadron is further divided into flights.

CAREER MOVES

How can you prepare for a career in the US Air Force?

★ Take English, math, science, and engineering classes to prepare for the ASVAB test, which all military recruits must take.

★ Get into good physical shape through running or other cardiovascular exercise, weight training, and flexibility training.

★ Talk to recruiters or other US Air Force personnel about what to expect in the air force.

★ Research potential jobs in the air force.

IMPACT ON SOCIETY

Since its formation, the US Air Force has patrolled the skies and protected the United States, its citizens, and its allies. To do this, the air force trains, maintains, and equips the country's military aircraft and aviation personnel. It prepares them to fight and win wars and protect the country against threats. In addition to defending the United States, the air force also helps when disasters strike around the world and sends airmen, aircraft, and supplies to provide humanitarian aid to those in need.

QUOTE

"I love my job, and I tell people that if they weren't paying me, I would pay them because I absolutely love it. We jump out into the middle of the ocean, inflate boats, and drive them to shore. I have had some amazing experiences."

—Ben Domian, air force reserve SERE specialist

GLOSSARY

avionics
Electronics used in aircraft.

cadet
A student or trainee in the armed services.

civilian
A person not serving in the armed forces.

coalition
A collection of groups or people that have joined together for a common purpose.

commando
A highly skilled soldier trained to carry out raids separately from a main military force.

court-martial
A trial for members of the military in a special court because they are accused of breaking military law.

cruise missile
A low-flying missile that is guided to its target by an onboard computer.

cyberspace
The online world of the internet and computer networks.

deploy
To spread out strategically; to send into battle.

forage
To search for food or supplies.

GED
A General Education Development certificate, which proves a person's high school–level education.

humanitarian
Concerned with relieving human suffering.

intelligence
Information that is of military or political value.

intercontinental ballistic missile
A weapon that flies under its own power and uses a guidance system to strike a specific target at long range.

marksmanship
The skill of shooting at and hitting a target.

phishing
An internet scam used by thieves to steal information and money from victims, or to install malicious software.

reconnaissance
An exploration of an area to gather information about the activity of military forces.

recruit
A person who has decided to enlist in the military.

sanction
An action taken to enforce the law.

satellite
An object, often human-made, orbiting a planet or moon that is larger than itself.

surveillance
Close observation or watch kept over something or someone.

ADDITIONAL RESOURCES

Selected Bibliography

"Careers." *US Air Force*, n.d., airforce.com/careers. Accessed 7 Jan. 2020.

Dolezal, Cliffton. "A Day in the Life of a Pilot." *Little Rock Air Force Base*, 4 Sept. 2014, littlerock.af.mil. Accessed 7 Jan. 2020.

Lange, Katie. "Military Units: How Each Service Is Organized." *DODLive*, 17 May 2017, dodlive.mil. Accessed 11 Dec. 2019.

Further Readings

Henzel, Cynthia Kennedy. *US Army*. Abdo, 2021.

Lusted, Marcia Amidon. *US Special Operations Forces*. Abdo, 2021.

Mitchell, P. P. *Join the Air Force*. Gareth Stevens Publishing, 2018.

Online Resources

To learn more about the US Air Force, please visit **abdobooklinks.com** or scan this QR code. These links are routinely monitored and updated to provide the most current information available.

More Information

For more information on this subject, contact or visit the following organizations:

Air Force Space & Missile Museum
191 Museum Circle
Patrick AFB, FL 32925
321-853-3246
afspacemuseum.org

This museum features historical exhibits and artifacts related to the US Air Force's role in the United States space program. It includes several rockets, missiles, and related space equipment.

National Museum of the United States Air Force
1100 Spaatz St.
Wright-Patterson AFB, Ohio 45433
937-255-3286
nationalmuseum.af.mil

This museum, located at Wright-Patterson Air Force Base near Dayton, Ohio, features historical artifacts and exhibits related to the US Air Force.

US Air Force Academy Barry Goldwater Visitors Center
2346 Academy Dr.
United States Air Force Academy, CO 80840
719-333-2025
www.usafa.edu/academics/facilities/visitors-center

Named after former Arizona Senator Barry Goldwater, who supported the US Air Force Academy, this center offers visitors information, including films and exhibits, about the academy's history and life for air force cadets.

SOURCE NOTES

CHAPTER 1. FLYING BEHIND ENEMY LINES

1. "Air Force." *Today's Military*, n.d., todaysmilitary.com. Accessed 2 Mar. 2020.

2. "Air National Guard." *Today's Military*, n.d., todaysmilitary.com. Accessed 2 Mar. 2020.

CHAPTER 2. THE HISTORY OF THE US AIR FORCE

1. "Air Force History." *Military.com*, n.d., military.com. Accessed 2 Mar. 2020.

2. Ed Grabianowski. "How the US Air Force Works." *How Stuff Works*, 5 Mar. 2007, science.howstuffworks.com. Accessed 2 Mar. 2020.

3. "Air Force History."

4. Blake Stilwell. "These Are the 7 Finest Moments in Air Force History." *We Are the Mighty*, 20 Sept. 2018, wearethemighty.com. Accessed 2 Mar. 2020.

5. "Air Force History."

6. Andrew Glass. "This Day in Politics: LBJ Approves 'Operation Rolling Thunder,' Feb. 13, 1965." *Politico*, 13 Feb. 2019, politico.com. Accessed 3 Mar. 2020.

7. Stilwell, "These Are the 7 Finest Moments in Air Force History."

CHAPTER 3. THE US AIR FORCE TODAY

1. "Demographics." *Air Force's Personnel Center*, n.d., afpc.af.mil. Accessed 2 Mar. 2020.

2. "Education at a Different Altitude." *US Air Force Academy*, n.d., usafa.edu. Accessed 2 Mar. 2020.

3. "Air Force Organization 101." *Aerospace Security*, 6 Dec. 2017, aerospace.csis.org. Accessed 2 Mar. 2020.

4. "Air Force Organization 101."

5. Ed Grabianowski. "How the US Air Force Works." *How Stuff Works*, 5 Mar. 2007, science.howstuffworks.com. Accessed 2 Mar. 2020.

6. "US Air Force." *Heritage.org*, 30 Oct. 2019, heritage.org. Accessed 2 Mar. 2020.

7. "What to Expect: Every Choice Brings Change." *US Air Force*, n.d., airforce.com. Accessed 2 Mar. 2020.

8. Mac McClelland. "Sky Warriors: A Day in the Life of Women in the US Air Force." *Allure*, 21 June 2018, allure.com. Accessed 2 Mar. 2020.

CHAPTER 4. PILOT

1. Cliffton Dolezal. "A Day in the Life of a Pilot." *Little Rock Air Force Base*, 4 Sept. 2014, littlerock.af.mil. Accessed 2 Mar. 2020.

2. Dolezal, "A Day in the Life of a Pilot."

3. Fernando Vergara, Gisela Salomon, and Fabiola Sanchez. "Air Force C-17s Bring Tons of Humanitarian Aid for Venezuela." *Air Force Times*, 17 Feb. 2019, airforcetimes.com. Accessed 2 Mar. 2020.

4. Nick Bente. "5 Things No One Told Me about Being a Pilot in the Air Force." *Task & Purpose*, 12 Dec. 2014, taskandpurpose.com. Accessed 2 Mar. 2020.

5. "ROTC Programs." *Today's Military*, n.d., todaysmilitary.com. Accessed 2 Mar. 2020.

6. "Air Force Announces Fiscal Year 2019 Aviation Bonuses." *US Air Force*, 23 Jan. 2019, af.mil. Accessed 2 Mar. 2020.

CHAPTER 5. TACTICAL AIRCRAFT MAINTENANCE SPECIALIST

1. Alexander Martinez. "A Day in the Life: C-17 Crew Chief." *15th Wing*, 15 Oct. 2014, 15wing.af.mil. Accessed 2 Mar. 2020.

2. Martinez, "A Day in the Life: C-17 Crew Chief."

3. Jordyn Fetter. "Face of Defense: Crew Chief Keeps Air Force Jets Flying." *US Department of Defense*, 14 July 2016, defense.gov. Accessed 2 Mar. 2020.

4. Caitlin Foster. "Military Pay: This Is How Much US Troops Are Paid According to Their Rank." *Business Insider*, 15 Feb. 2019, businessinsider.com. Accessed 2 Mar. 2020.

CHAPTER 6. SURVIVAL, EVASION, RESISTANCE, AND ESCAPE SPECIALIST

1. Brian Everstine. "SERE Specialists Working to Expand Their Ranks, Improve Recruitment." *Air Force Magazine*, 1 Jan. 2019, airforcemag.com. Accessed 2 Mar. 2020.

2. Kayshel Trudell. "SERE Specialists Showcase Training for Recruiters." *Joint Base San Antonio*, 17 June 2019, jbsa.mil. Accessed 2 Mar. 2020.

3. Clint Carter. "Inside America's Toughest Survival School." *Men's Journal*, n.d., mensjournal.com. Accessed 2 Mar. 2020.

4. "Ben Domian: A Life Influenced by Family and Lance Sijan." *Milwaukee Independent*, 5 May 2017, milwaukeeindependent.com. Accessed 2 Mar. 2020.

5. SERE Recruiting Liaison Office. "SERE Specialist Training Orientation Course (SST-OC)." *US Air Force*, 11 Apr. 2017, gosere.af.mil. Accessed 2 Mar. 2020.

6. "SERE." *US Air Force*, n.d., gosere.af.mil. Accessed 2 Mar. 2020.

7. Carter, "Inside America's Toughest Survival School."

CHAPTER 7. WEATHER OFFICER

1. Devin James. "Weather You Know It or Knot: Air Force Staff Weather Officers Play Crucial Role in Army Planning." *US Army*, 5 Feb. 2014, army.mil. Accessed 2 Mar. 2020.

2. "Meteorology." *National Geographic*, 12 Jan. 2012, nationalgeographic.org. Accessed 2 Mar. 2020.

3. James, "Weather You Know It or Knot."

4. James, "Weather You Know It or Knot."

5. "Internships Help Ready the Next Generation of US Air Force Weather Officers." *Virginia Tech*, 3 Apr. 2019, vtnews.vt.edu. Accessed 2 Mar. 2020.

6. Ed Grabianowski. "How the US Air Force Works." *How Stuff Works*, 5 Mar. 2007, science.howstuffworks.com. Accessed 2 Mar. 2020.

CHAPTER 8. CYBER SURETY SPECIALIST

1. Stephen Losey. "Go Cyber: Airmen Can Earn Cash and Promotions, Get Set Up for Civilian Life." *Air Force Times*, 10 Oct. 2017, airforcetimes.com. Accessed 2 Mar. 2020.

2. Amanda Morris. "Gone Phishing: Cyber Surety Technicians Keep Watchful Eye." *Andersen Air Force Base*, 11 June 2015, andersen.af.mil. Accessed 2 Mar. 2020.

3. Morris, "Gone Phishing."

4. Losey, "Go Cyber."

5. "Monthly Rates of Basic Pay (Enlisted)." *Defense Finance and Accounting Service*, 1 Jan. 2020, dfas.mil. Accessed 2 Mar. 2020.

CHAPTER 9. PARALEGAL SPECIALIST

1. Chris McCann. "BER Senior Master Sergeant Recognized as One of Air Force's 12 Best." *Joint Base Elmendorf-Richardson*, 15 Aug. 2019, jber.jb.mil. Accessed 2 Mar. 2020.

2. McCann, "BER Senior Master Sergeant."

3. McCann, "BER Senior Master Sergeant."

4. McCann, "BER Senior Master Sergeant."

5. David Smith. "'My Transition' #41: Donte Anderson—Military Legal Operations Manager to Current Job Seeker." *Medium*, 30 Jan. 2018, medium.com. Accessed 2 Mar. 2020.

INDEX

ABOUT THE AUTHOR

Carla Mooney

Carla Mooney is the author of many books for young adults and children. She lives in Pittsburgh, Pennsylvania, with her husband and three children.